INCREDIBLE MACHINES!

The Physics of **Simple Machines**

Written by Laura Perdew

www.worldbook.com

Co-published by agreement between Shi Tu Hui and World Book, Inc.

Shi Tu Hui
Room 1807, Block 1,
#3 West Dawang Road
Chaoyang District, Beijing 100025
P.R. China

World Book, Inc.
180 North LaSalle Street
Suite 900
Chicago, Illinois 60601
USA

Library of Congress Control Number: 2024947056

Aha! Academy: Physics
ISBN: 978-0-7166-7144-2 (set, hard cover)

Incredible Machines! The Physics of Simple Machines
ISBN: 978-0-7166-7149-7 (hard cover)
ISBN: 978-0-7166-7169-5 (e-book)
ISBN: 978-0-7166-7159-6 (soft cover)

Printed in India by Replika Press PVT LTD, Haryana, India
1st printing January 2025

Staff

Editorial

Vice President
Tom Evans

Editorial Project Coordinator
Kaile Kilner

Senior Curriculum Designer
Caroline Davidson

Proofreader
Nathalie Strassheim

Graphics and Design

Senior Visual
Communications Designer
Melanie Bender

Designer
Shannon Hagman

Digital Asset Specialist
Rosalia Bledsoe

Written by Laura Perdew
Advised by Anthony Stewart

Developed with World Book by
Red Line Editorial

Acknowledgments

The publishers gratefully acknowledge the following sources for photography. All illustrations were prepared by WORLD BOOK unless otherwise noted.

Cover: Brocreative/Shutterstock; FooTToo/Shutterstock; Olivkairishka/Shutterstock; Aleksandr Semenov, Shutterstock; theshots/Shutterstock

三猎 (licensed under CC BY 4.0) 21; Graphic House/Archive Photos/Getty Images 39; stevecoleimages/Getty Images 7; Public Domain (Heritage Auctions/Collier's Weekly) 43; Shutterstock 3, 4, 5, 6, 7, 8, 9, 10, 11, 12, 13, 14, 15, 16, 17, 18, 19, 20, 21, 22, 23, 24, 25, 26, 27, 28, 29, 30, 31, 32, 33, 34, 35, 36, 37, 38, 39, 40, 41, 42, 43, 44, 45, 46, 47, 48

There is a glossary of terms on page 48. Terms defined in the glossary are in type that looks like ***this*** on their first appearance on any spread (two facing pages).

Contents

Introduction

The rider jumps on a skateboard and drops into the bowl. They ride up one side, perform a trick midair, then land and roll back down into the bowl. The momentum sends them back up the other side to do another trick. This may seem like play. But it's really ***work!***

All six simple machines are in an airplane!

Work occurs anytime someone uses ***force*** to move an object some distance. At the skateboard park, riders use structures in the park and their board to move themselves and to do tricks. There's something else at play here too—simple machines.

Simple machines are tools with few or no moving parts. And while they are simple, they are mighty! Each of the six simple machines makes work easier. Putting two or more simple machines together creates a compound machine that makes work even easier.

Read on to find out what else you can do with the help of simple machines!

INCREDIBLE INCLINED PLANES

You need to move heavy props onto a stage for the school play. Lifting them straight up with your arms would be difficult. But you could use an inclined plane, or a ramp.

An inclined plane is a flat surface with one end higher than the other. That's it. Simple. Inclined planes help move such objects as props (a ***load***) up or down gradually.

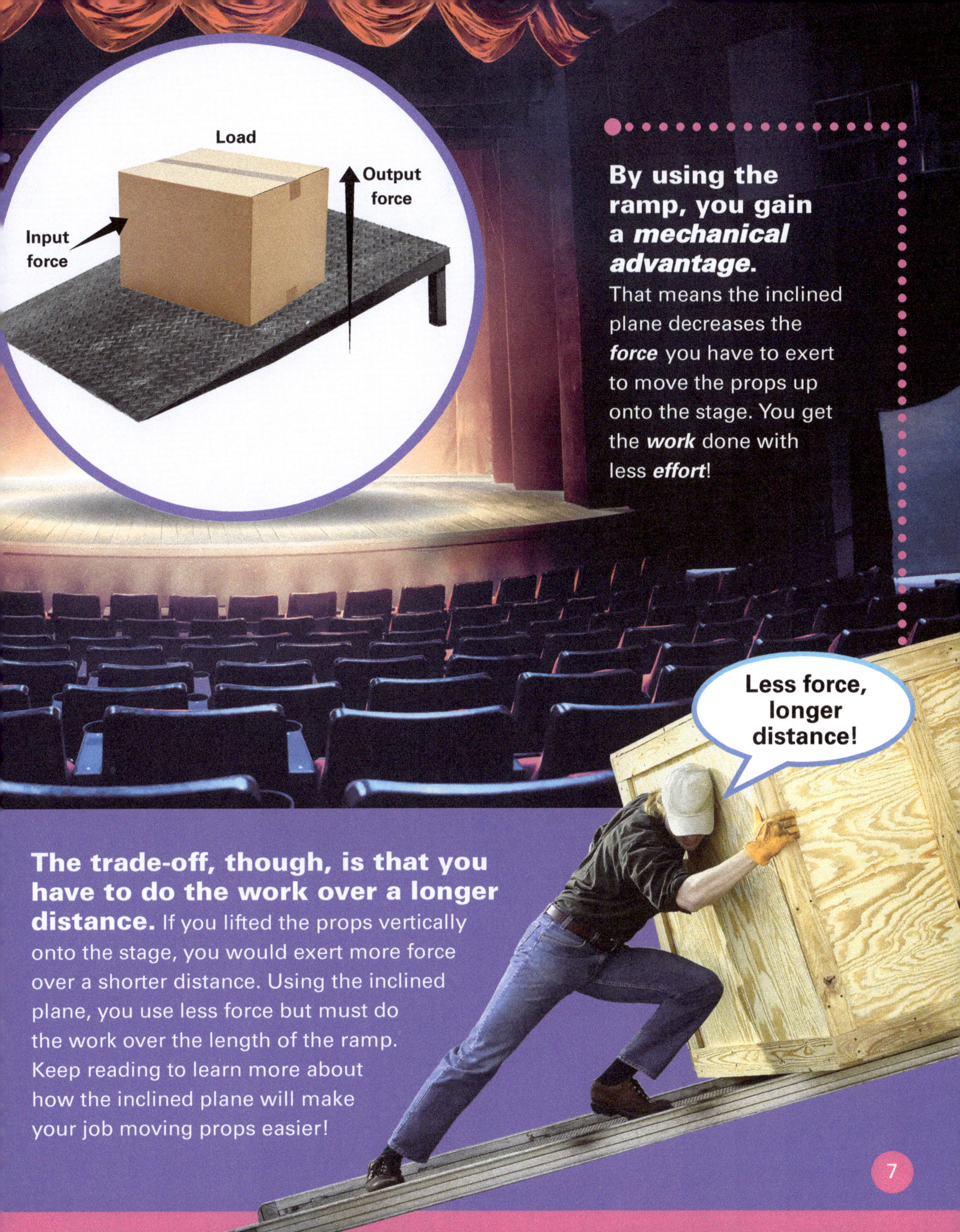

By using the ramp, you gain a *mechanical advantage*. That means the inclined plane decreases the ***force*** you have to exert to move the props up onto the stage. You get the ***work*** done with less ***effort***!

The trade-off, though, is that you have to do the work over a longer distance. If you lifted the props vertically onto the stage, you would exert more force over a shorter distance. Using the inclined plane, you use less force but must do the work over the length of the ramp. Keep reading to learn more about how the inclined plane will make your job moving props easier!

Simple **planes**

The *output force* is the end result. In other words, it is the force exerted by the simple machine (the inclined plane) to move the props vertically to the higher point.

Something else to consider with an inclined plane is how steep it is. If you use a steep ramp with a large angle, you'll need more force to move those props up onto the stage than if you use a longer ramp that is less steep.

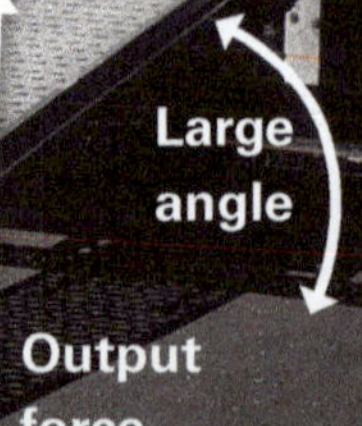

Did you notice that when you use an inclined plane, you also change the direction of the *force*? You do! To get those props onstage, you and your muscles supply the ***input force*** diagonally to drag, push, or carry the ***load***. Input force is the force applied to the machine.

Inclined planes are at work all around you. Wheelchair ramps, roads, and moving truck ramps are all examples of inclined planes that help us move a load upward. Stairs and ladders are also a type of inclined plane. Look around—what other inclined planes do you see?

The **Great Pyramid of Giza**, in Egypt, was built more than 4,000 years ago without any modern technology. Can you imagine trying to move massive stone blocks into place? Just one of them weighed more than 2 tons! Ancient engineers devised ramps that workers used to haul the blocks up into position.

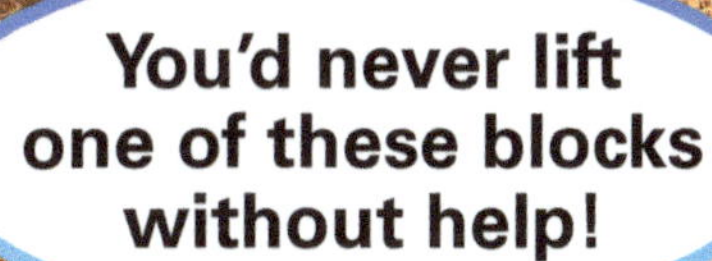

What goes up **must come down**

If you use the ramp again, you can move all the props down slowly. Because ***gravity*** is a ***force*** that pulls objects downward, you want to use a ramp. It decreases the speed so your ***load*** doesn't move too fast. This is the same reason you use stairs to move down from the second floor of a building to the first floor rather than jumping.

The angle of the inclined plane also determines how fast something moves downward. Think about a playground slide. If the slide has a steep slope, you slide down faster. But as the angle of the slide decreases, so will your speed.

DID YOU KNOW?

The beds of dump trucks work like ramps too. They tilt up to unload material, creating an inclined plane to let the materials slide out. That's much easier than moving it by hand!

After the play, you don't want to drop the fog machine off the edge of the stage. It might break. And what good is a broken fog machine? Inclined planes are also used to move objects downward gradually.

Ski slopes are another example of an inclined plane. If you are a skilled skier, you may want a steep slope to go faster. But if you are new to skiing, you'll want a gentler slope.

CLEVER LEVERS

Imagine you are on one end of a seesaw and your friend is on the other. As you go up, they go down. Without the seesaw, you probably wouldn't be able to lift your friend into the air. But because the seesaw is a simple machine, you can!

Input force

The seesaw is a type of lever. A lever is simply a bar, or arm, that sits on a point called a ***fulcrum***. The lever moves on the fulcrum. When ***force*** is exerted on the lever, it moves a ***load***. Such levers as seesaws, catapults, and balance scales are used to lift, launch, and balance objects.

Bar

Fulcrum

Your seesaw is a first-class lever, with the fulcrum in the middle. When you are up in the air, you exert a downward force on the bar. Your friend on the other end is the load. As you exert a downward force, the ***output force*** moves your friend upward.

Ancient Greek mathematician **Archimedes** once said, "Give me a lever and a place to stand and I shall move the world." Archimedes wasn't bragging about how strong he was. He was describing the ***mechanical advantage*** of a lever. If Archimedes had a very, very, very, very long lever and the fulcrum were placed close to Earth, his input force would be greatly multiplied. The output force would be enough to move the planet.

Now imagine that your friend is actually a friendly hippopotamus who weighs much more than you do. One way to lift your friend would be for you (the ***input force***) to move farther away from the fulcrum. By doing that, the ***work*** is done over a longer distance. The other option is for your friend (the load) to move closer to the fulcrum so the friend moves a shorter distance.

Other examples of a first-class lever include a crowbar, piano pedals, oars on a boat, and pliers.

Caps off!

A bottle opener is a second-class lever. Its ***load*** is between the ***fulcrum*** on one end and the ***force*** on the other. The fulcrum is where the tip of the opener sits on the bottle. You apply an upward ***input force*** to the bottle opener. By pulling up on the opener, you are able to lift off the bottle cap, or load.

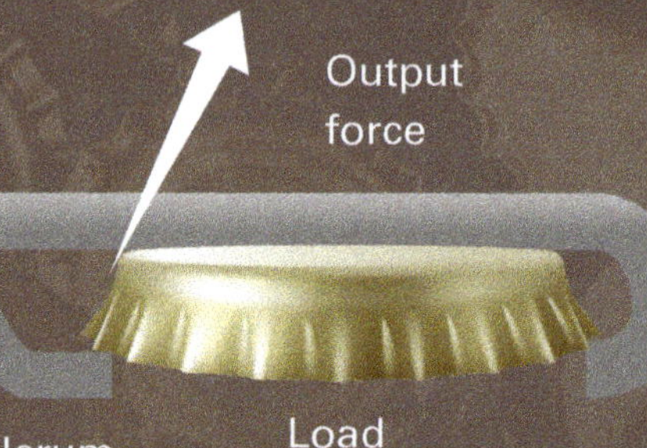

A second-class lever reduces the force needed to move a load, but it is applied over a longer distance (the length of the lever). When you remove a bottle cap, the distance is the length of the bottle opener, or arm. Your input and the ***output forces*** will move in the same direction when using a second-class lever.

How would you get the bottle cap off a soda bottle?

Unless it had a screw top, you probably wouldn't be able to pry the cap off with just your fingers. You'd use a lever!

Second-class levers are all around you.

Doors

Every time you open or close a door, you use a lever. The hinges are the fulcrum, the doorknob is where you apply the force, and the door is the load.

Nutcrackers

Where the two arms of a nutcracker meet is the fulcrum. You apply the force on the arms of the device, and the nut is the load.

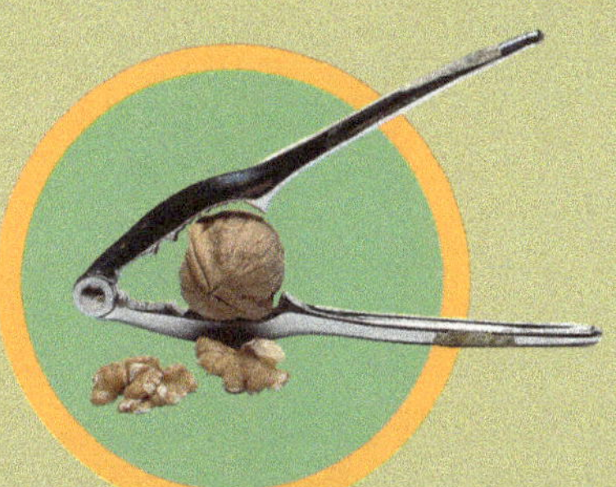

Your body

The human body is even a second-class lever! When you stand on tiptoes, your toes are the fulcrum, and your calf muscles apply the force; the load is your body.

Anybody seen a beehive?

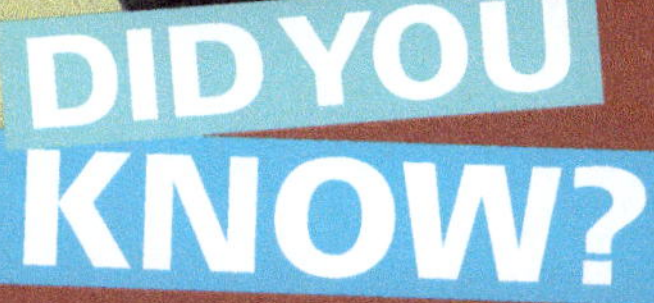

Some animals use levers too! Crows and chimpanzees use sticks as levers to pry food (the load!) out of a hole. Chimpanzees are known to use a stick as a lever to widen the opening of a beehive.

Batter up!

Baseball bats, hockey sticks, and tennis rackets are all third-class levers! They make it possible for you to move a ball or puck much faster than you could with just your hands or feet.

A third-class lever applies the *force* in the middle. The fulcrum is on one end, and the load is on the other. The ***input*** and ***output forces*** move in the same direction when using a third-class lever.

The third-class lever does not reduce the amount of force you need to do *work*. In fact, more effort is needed to move a load. However, third-class levers change the speed at which a load moves. The lever, a bat, allows you to send the ball past the opposing team!

Did you know that a baseball bat is a lever? Grab one and prepare to swing. The spot where you grip the bat is the ***fulcrum***. The ***effort*** is applied above your hands. Swing and hit the ***load***—SMACK! The ball soars!

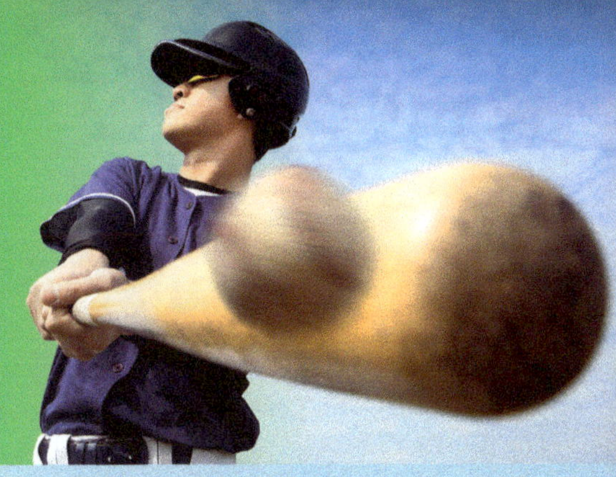

Levers help us collect leaves, hammer nails, and sweep. Rakes, hammers, and brooms are third-class levers.

We use third-class levers in the kitchen too. Chopsticks and tongs are third-class levers.

Your arm is a third-class lever. Your elbow is the fulcrum on one end, your forearm applies the force, and your hand holds an object you want to lift.

CURIOUS CONNECTIONS

Doctors and engineers use levers when building prosthetics, or artificial body parts. Adjusting the toe lever or heel lever on a prosthetic leg changes how a user walks. Arm prosthetics also use levers and gears to improve how an arm works and moves. The levers in prosthetics mimic those in the human body.

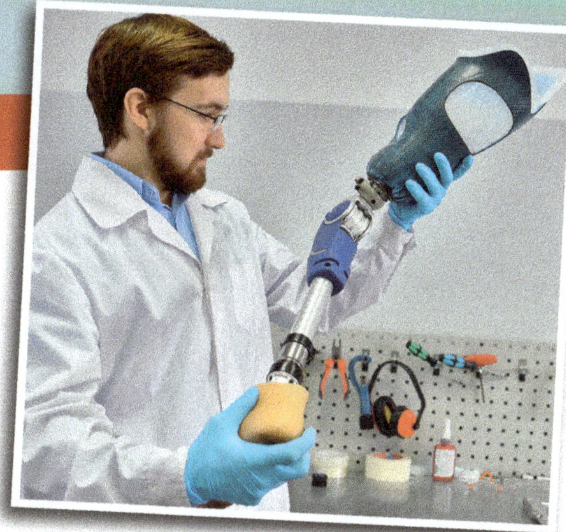

3 WONDERFUL WEDGES

When you prepare food, you don't just throw whole vegetables into a pan or a bowl. You want them in bite-sized pieces. Pulling vegetables apart by hand would take a lot of time and *effort*. A knife is a type of simple machine called a wedge.

Wedges are wide on one end and narrow to a thin edge or point on the other. When you use a wedge, the ***output force*** is greater than the ***input force***. In addition, a wedge changes a pushing ***force*** into a splitting force. That's what happens when you use a knife—you push down on the blade, and the pieces of vegetable move in two different directions.

Different types of *work* call for different kinds of wedges. When chopping vegetables or wood, you might use a short, fat wedge. But you would need to apply a strong force. If you want the work to be easier, choose a long, thin wedge, like a knife! That decreases the force you have to apply, as the work is done over a longer distance. The longer and thinner the wedge, the greater the ***mechanical advantage***.

DID YOU KNOW?

Bite into an apple. The wedge of your front teeth cuts the food apart. Your front teeth are wider along your gums and narrower at the tips. Your upper and lower front teeth work together to pull off a piece of the apple.

Compounding the work of wedges

These machines multiply the *mechanical advantage* provided by the simple machines that make up the complex one. For us, compound machines make ***work*** even easier and save time!

So what happens when you combine a wedge and a lever?
You might get a stapler! A stapler combines a lever (the stapler arm) and a wedge (the staples). The lever helps you apply a greater ***input force*** and push the wedge into the paper.

Wedges are great tools, but people have found ways to make them even better! Sometimes two or more simple machines, like a wedge and a lever, are combined to create a compound machine.

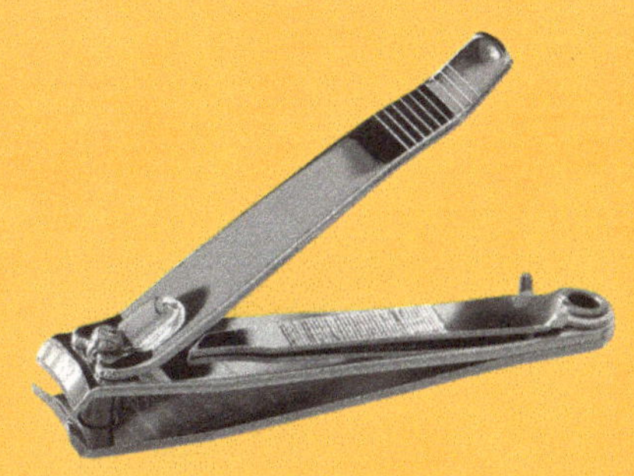

Other wedge-and-lever combos include nail clippers and axes.

We're thousands of years old!

TECH TIME

The wonderful tool we call scissors was invented more than 3,000 years ago in the Middle East. These first scissors looked a bit like a large set of metal tweezers or tongs, with two blades facing each other. When the blades were squeezed together, they cut like scissors. In A.D. 100 the Romans created the pivoted scissors. These are similar to what we use today. Both designs are compound machines of two levers and two wedges.

WILD ABOUT WHEELS

Think about bringing home a heavy *load* of groceries. You could carry it. If it's too heavy, you could use a box and push it along the ground. But in doing that you have both ***friction*** and ***gravity*** working against you, making your ***work*** harder.

Put those groceries in a cart with wheels, though, and you're off! The wheel and axle form another simple machine that makes doing work easier.

The wheel and axle consist of a disc (the wheel) with a rod of a smaller *diameter* (the axle) fixed to its center. When ***force*** is applied to one or the other, it causes the wheel and axle to rotate together, changing a linear force into a rotational force. The greatest advantage of a wheel and axle is that this simple machine reduces friction. This makes it easier to move all your groceries at once!

The **first wheels** ever made were used to make pottery around 3500 B.C. in Mesopotamia. They were made of wood or stone, and people rotated them manually. It wasn't until several hundred years later that the wheel was used on vehicles. Archaeological evidence shows that carts with wheels were used in parts of Europe, Asia, and the Middle East.

Want to know another way to make moving your load easier? Look for the cart with the biggest wheels. A wheel's radius is like a lever! And as with a lever, the longer the radius, the greater the ***mechanical advantage***.

The source of **the force**

Wheels and axles can work in two different ways. One way is to apply ***force*** to the axle, which then makes the wheel turn. That's how a Ferris wheel works. The motor turns the axle, which turns the larger wheel, and around you go!

The short *input force* on the axle of the Ferris wheel makes the wheel where you ride move a longer distance. Applying force to the axle like this does not provide a ***mechanical advantage***, but it does reduce ***friction*** and increase speed, making the ride smooth and fun. Here are some other wheels and axles in which the force is applied to the axle: car wheels, revolving doors, a merry-go-round, and a rolling pin.

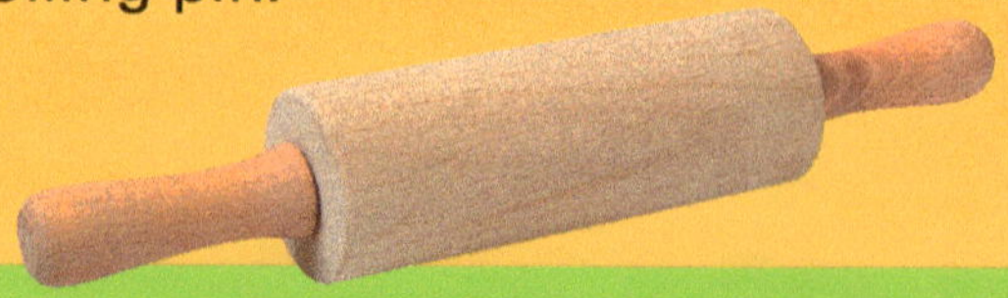

You're at the top of a Ferris wheel, checking out the view of the local scene. But how does the wheel and axle in this thing work?

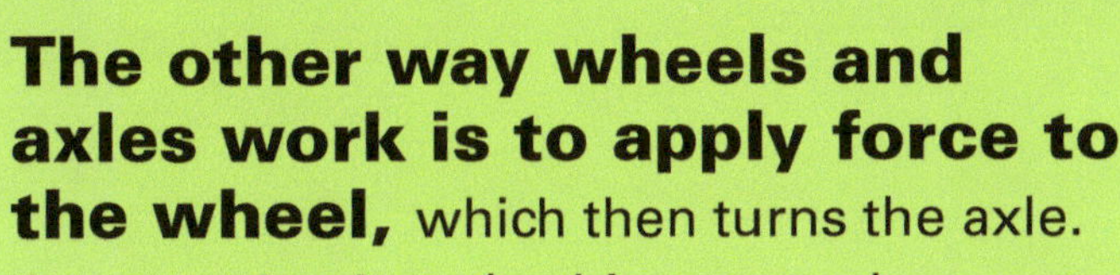

The other way wheels and axles work is to apply force to the wheel, which then turns the axle. A waterwheel works this way—when water travels through a channel, falls, and hits the wheel, it makes the wheel spin. When the wheel spins, it turns the axle. These are some other examples of this type of wheel-and-axle system: doorknobs, screwdrivers, skateboards, and car steering wheels.

CAREER CORNER

Playgrounds include many different simple machines. There might be a seesaw and a merry-go-round. Ramps, stairs, and ladders get you into the structures. And straight or swirly slides help you get down. Some playgrounds even have a zip line (which is a pulley system). Each playground is designed by a playground architect! These designers need a background in simple machines, engineering, and landscape architecture. They must also put the playground elements together in creative ways.

Wheels for food, fans, and fun

Electric fans apply force to the axle to turn the hub where the blades of the fan are attached. The blades, though, are wedges that cut through the air and create a cool breeze.

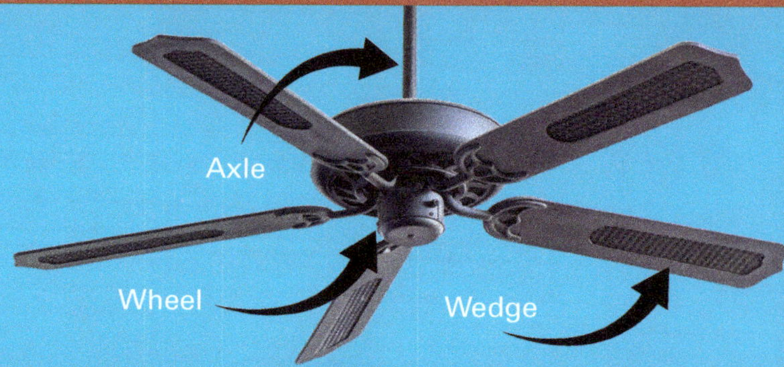

Even slicing a pizza is considered *work!* But if you have a pizza cutter, your job is not as difficult and you can eat sooner. It combines a lever (the handle), a wheel and axle, and a wedge (the sharp edge of the wheel that does the cutting).

CAREER CORNER

Wind-energy engineers design windmills called turbines to generate clean electricity. The heart of this technology is a simple machine—the wheel and axle. When wind (the ***input force***) hits the blades, the turbine translates that linear force into a rotational force that generates energy. Each blade is a wedge, allowing it to cut through the air and spin more easily.

If you're gardening, how would you move a pile of rocks out of the garden? A wheelbarrow is made up of multiple simple machines. ***Force*** is applied to the wheel. This tool also has handles that allow you to lift the ***load***. Those handles are levers!

Let's move it!

Levers

Screw

Wheel and axle

Pulley

Obviously, bikes have two wheels and two axles. But they also include other simple machines. The brakes work like levers. When force is applied to the brakes, the brake pads clamp together on the wheel. The pedals work like levers too. You push down on the pedals, which applies force to the axle, to get the bike moving.

Screws are another type of simple machine. They connect different parts of the bicycle together. And the gears and chain work like a pulley system, which is another simple machine! You'll learn more about screws and pulleys in the coming chapters!

5

SPECTACULAR SCREWS

If you were building a loft in your room for your bed, what would you use to hold it together?

Screws! These simple machines will let you sleep tight without worrying if your loft will fall apart in the middle of the night.

Look closely at a screw. It has a central rod. But what do the threads wrapped around the screw look like? An inclined plane! A screw is simply an inclined plane spiraled around a central axis, the rod. Because of this, screws provide a high ***mechanical advantage***. It's a whole lot easier to turn screws into your loft frame than to try to push them straight in.

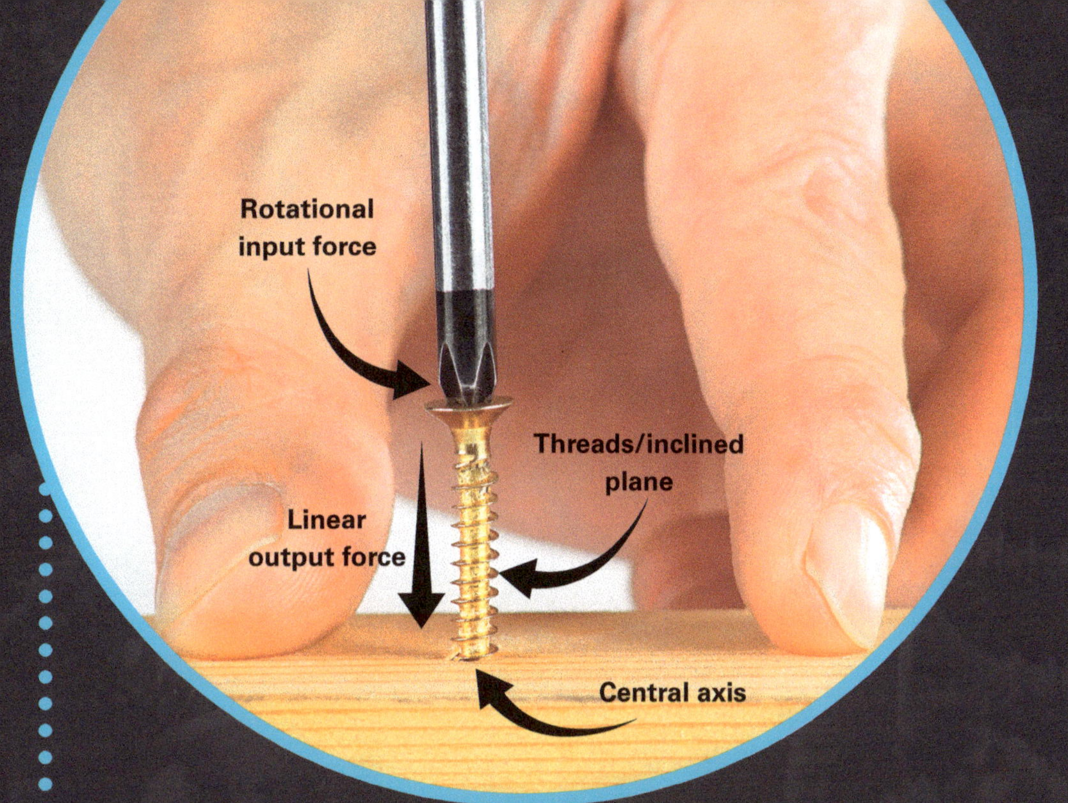

Screws change the direction of the *input force*. When the screwdriver and screws are turned, the input force is rotational. However, the ***output force*** is linear. As you build the loft, you turn the screws, which drives them straight into the wood to hold the pieces together tightly.

You may also notice that you have to do the work of turning the screw many times to move it into the wood a short distance. That's the trade-off with a screw—it's easier to move it into the wood, but it takes many turns.

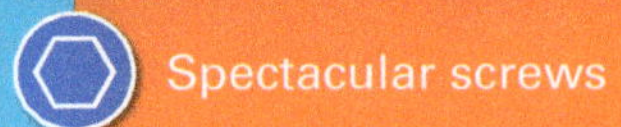

Holding it together

For all screws, how far apart the threads are determines the *mechanical advantage*. The closer the threads, the greater the mechanical advantage. It's like having a longer inclined plane over which the work is done. Because of that, threads that are closer together will hold materials together better.

How far apart the threads are and how many threads there are also determine how many times you have to turn the screw to tighten something. Go look at a sink with knobs that turn the hot and cold water on and off. These screws open and close valves, or devices that stop or start the flow of liquid, to deliver water through the faucet.

Screws hold many things together.

What makes screws different from one another, and how do they make ***work*** easier?

Twist open the lid on your water bottle and look inside it. There's a screw! There's another one on the neck of the bottle. Bottles and jars have screw threads that match their lids. When you twist the lid back on, the threads interlock. And the small ***input force*** you use creates a tight seal between the lid and the bottle so water doesn't leak into your bag!

DID YOU KNOW?

Have you ever heard the phrase "Righty tighty, lefty loosey?" This phrase refers to screws. It means that with most screws, turning to the right will tighten them. Turning to the left will loosen them.

A bolt works in combination with a nut that has matching threads on its inner circle. A nut and bolt are used as a fastener and interlock to hold something together.

Drilling with **screws and wedges**

An auger is like an oversized drill bit with two handles that form a T. The opposite end is pointed. It works as a wedge to split the ice.

The auger makes the *work* easier by reducing the *force* needed. It's also quicker! Augers are also used to put posts into the ground or plant trees. Another type of auger drills into drainpipes to remove clogs. The rotation of the auger bores, or punctures with a turning or twisting movement, into the material. It pulls the material up and out of the hole.

Do you know about ice fishing? That's when someone drills a hole in a frozen lake to be able to drop a line into the water below. They use a special kind of screw called an auger.

Many screws have a pointed tip, which is a wedge! That allows the screw to bore into another material with less ***effort***.

CURIOUS CONNECTIONS

AUTOMOTIVE TECHNOLOGY

Have you ever seen someone change a car tire? The car has to be raised off the ground. But no one can lift a car up using just their muscles! To lift a car, drivers and mechanics use a car jack that combines a lever and a screw. By turning the lever, the screw raises the jack and the car, turning a small ***input force*** over a long distance into a short, powerful force.

Screws to **lift and lower**

A circular staircase is an inclined plane that spirals around a central axis. As with an inclined plane, getting to the next floor in a building is easier using a circular staircase than scaling the walls or jumping. The amount of ***work*** (getting up to the next floor) stays the same, but the work is spread out over the distance of the circular stairs, so you input less ***force***.

Central axis

Inclined plane

Did you know that a circular staircase is a screw?

It is! Screws are used not only to hold things together. They also help to lift or lower objects.

Marble runs use screws too!

They send a marble down, down, down, and around. In this case, the use of the screw reduces the speed at which the marble descends, as opposed to having it fall straight down. Just as with an inclined plane, the steepness of the run will determine how fast the marble travels—the steeper the run, the faster the marble rolls.

AGRICULTURE

The screw has been used to irrigate, or water, crops for thousands of years. This irrigation method uses a hollow cylinder. Inside it is a screw. One end is placed in a body of water at an angle. Turning the screw scoops up water and moves it upward. If you continue to turn it, the water continues to move up and eventually comes out the top. This type of screw is still used to move water today.

6 POWERFUL PULLEYS

Imagine you're helping to lift a heavy bucket of water from a well.

How will you do that? You'll use another simple machine: a pulley!

Pulleys have a special wheel and axle.

The wheel has a groove on the outer edge. A rope sits in the groove, and the wheel guides its movement. As you pull down on the rope onstage, the curtain goes up.

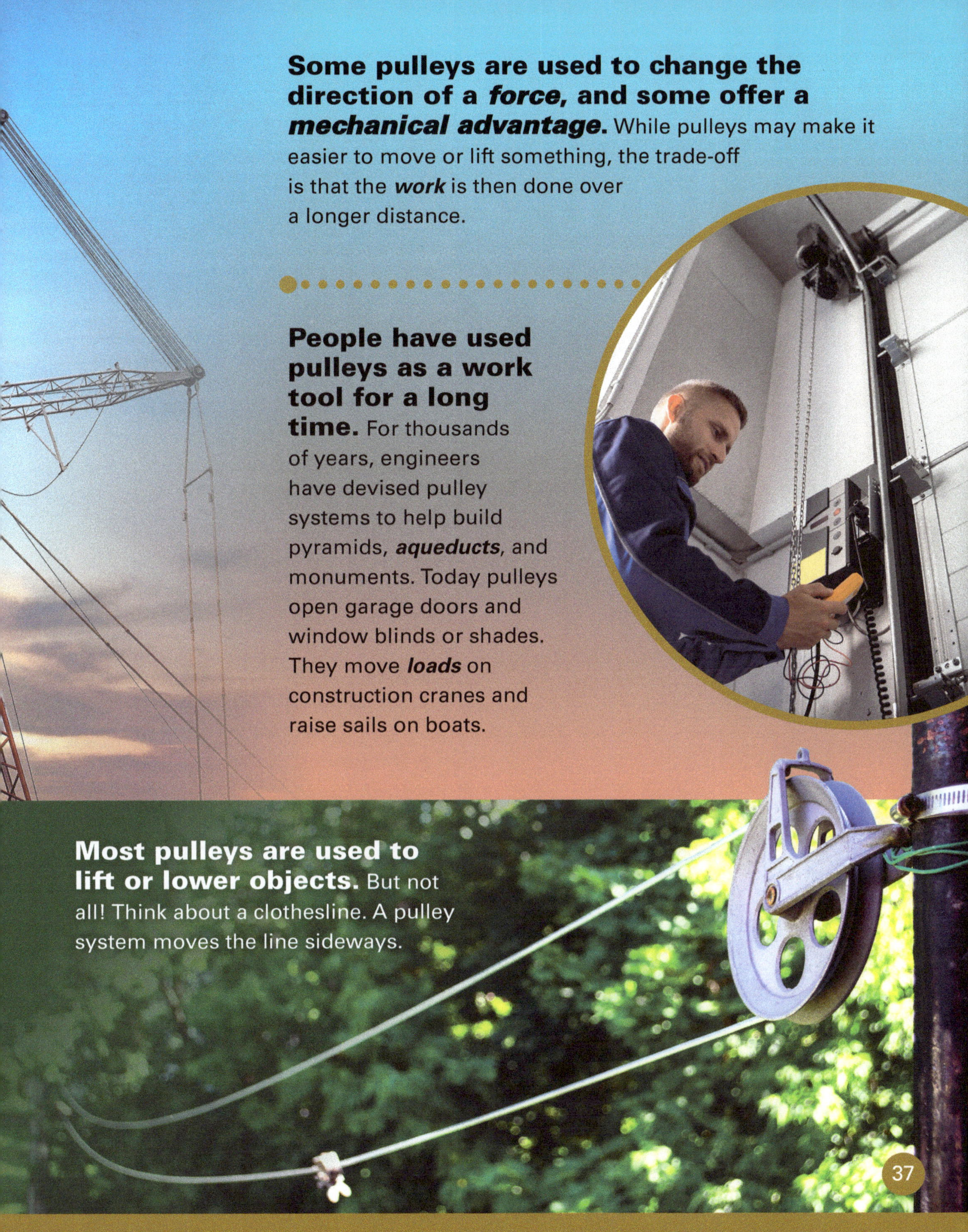

Some pulleys are used to change the direction of a *force*, and some offer a *mechanical advantage*. While pulleys may make it easier to move or lift something, the trade-off is that the ***work*** is then done over a longer distance.

People have used pulleys as a work tool for a long time. For thousands of years, engineers have devised pulley systems to help build pyramids, ***aqueducts***, and monuments. Today pulleys open garage doors and window blinds or shades. They move ***loads*** on construction cranes and raise sails on boats.

Most pulleys are used to lift or lower objects. But not all! Think about a clothesline. A pulley system moves the line sideways.

Decorating a tree house

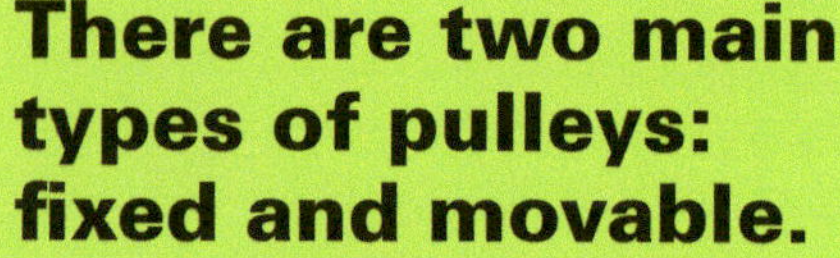

There are two main types of pulleys: fixed and movable. Fixed pulleys change the direction of a ***force***. They allow you to raise things to a higher location, as at the tree house.

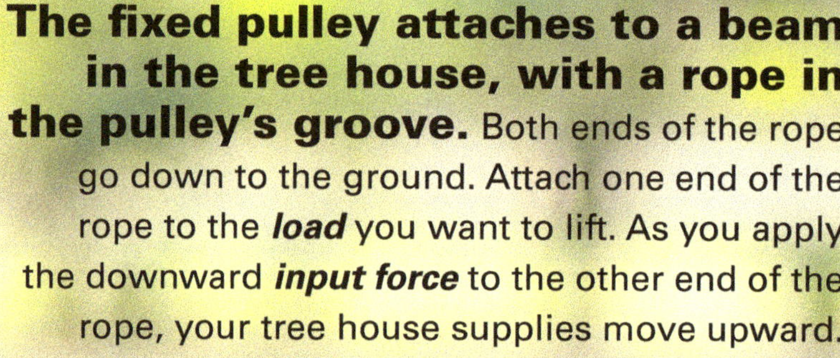

The fixed pulley attaches to a beam in the tree house, with a rope in the pulley's groove. Both ends of the rope go down to the ground. Attach one end of the rope to the ***load*** you want to lift. As you apply the downward ***input force*** to the other end of the rope, your tree house supplies move upward.

To get out of your tree house, you could just climb down the ladder. But a zip line would be much faster!

Your tree house high above the ground is the perfect hangout. But you need to get supplies up there. Some objects, such as the beanbag chairs, are too big to carry up the ladder. A fixed pulley will help!

Zip lines use the second type of pulley—the movable pulley. With this system, the load (you!) is attached to the pulley, and the pulley moves (or zips!) along the cable. ***Gravity*** applies the force to move you down along the zip line.

Unlike fixed pulleys, movable pulleys do offer a *mechanical advantage*. Construction cranes and some fitness equipment also use movable pulleys.

In the late 1800's, **Alice Bygrave** came up with an ingenious invention for women riding bicycles. At the time, women usually wore long skirts. That made riding a bicycle difficult and dangerous. Bygrave, a dressmaker in the United Kingdom, designed a skirt with a two-pulley system in it! One pulley attached to the front of the skirt and the other to the back. The wearer could adjust the hemline with the pulleys when riding a bike.

This skirt has pulleys!

Compound pulley power

When you do that, you create a compound pulley and increase the *mechanical advantage*. With each pulley you add, you reduce the amount of ***force*** needed to lift the couch. This system changes both the direction and the amount of force needed to do ***work***.

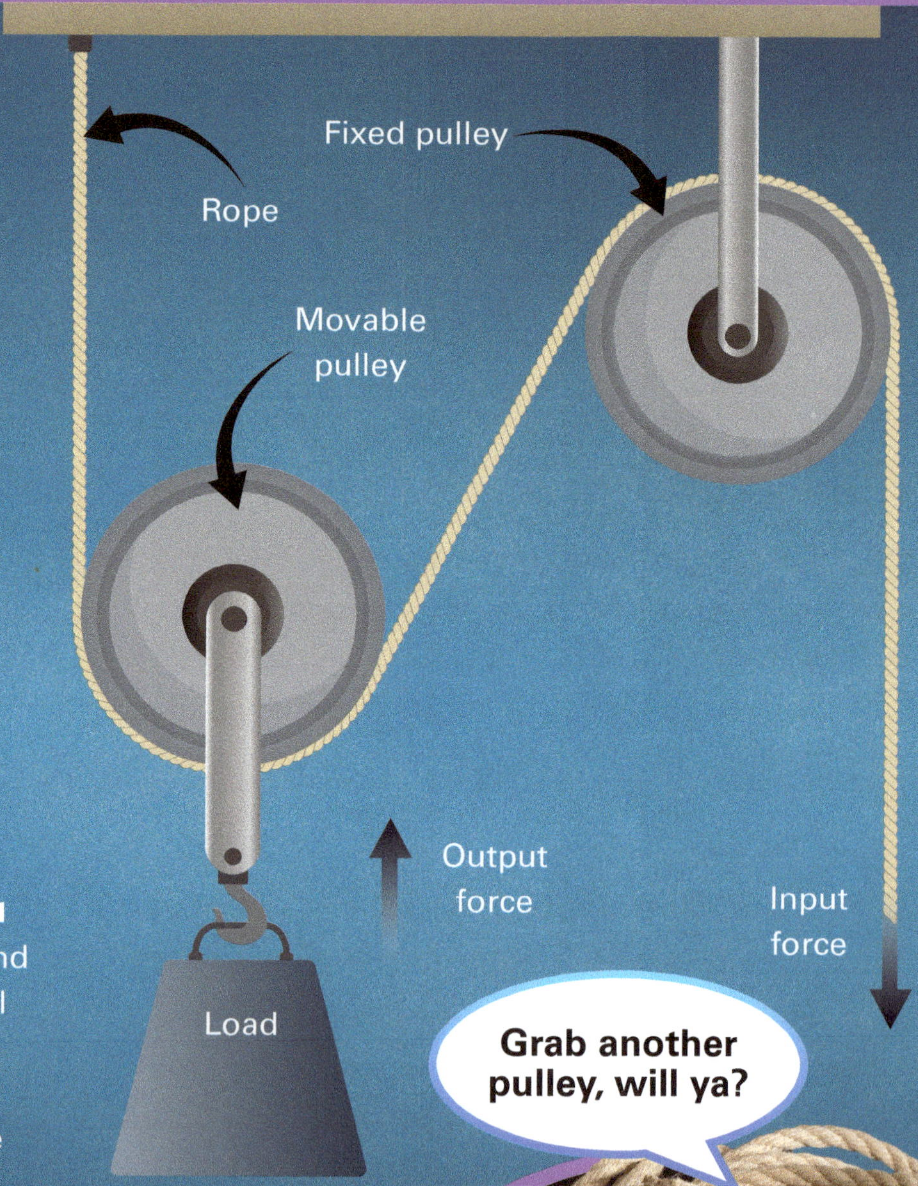

Every pulley you introduce cuts in half the force you need to do the work and increases the mechanical advantage. But you will also need twice as much rope, thus increasing the distance over which the work is done.

When moving things into your tree house, you might have a couch that's too heavy to move with just a fixed or a movable pulley. Don't worry. You'll still be able to get the couch into the tree house. Use both a fixed and a movable pulley together!

TECH TIME

Ever wondered what makes an elevator go up and down? Pulleys! The cars are attached to a thick metal cable that runs over a pulley. An electric motor provides the force. Some elevators use a ***counterweight*** to reduce the amount of force needed to lift the elevator car. It helps steady the elevator, and it makes it easier for the motor to raise and lower the car. Other elevator systems use compound pulleys to increase mechanical advantage.

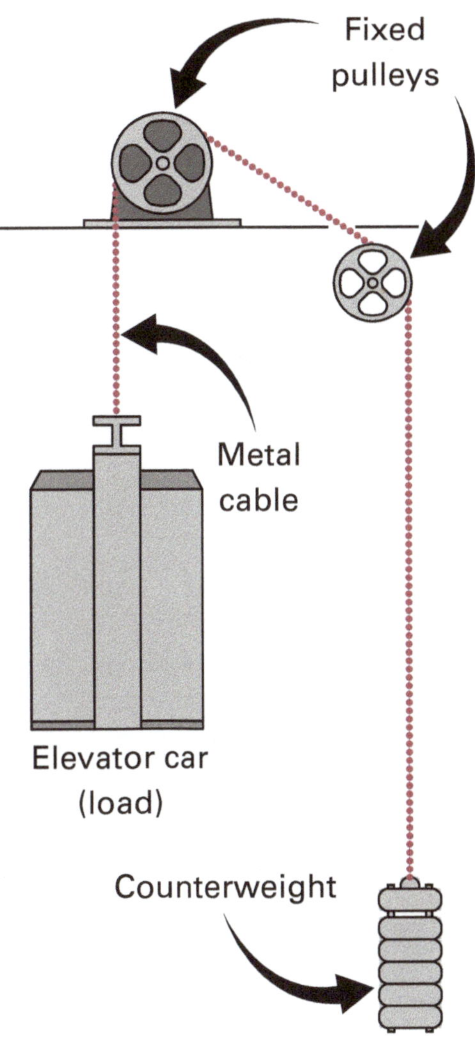

Simple machines at work and play

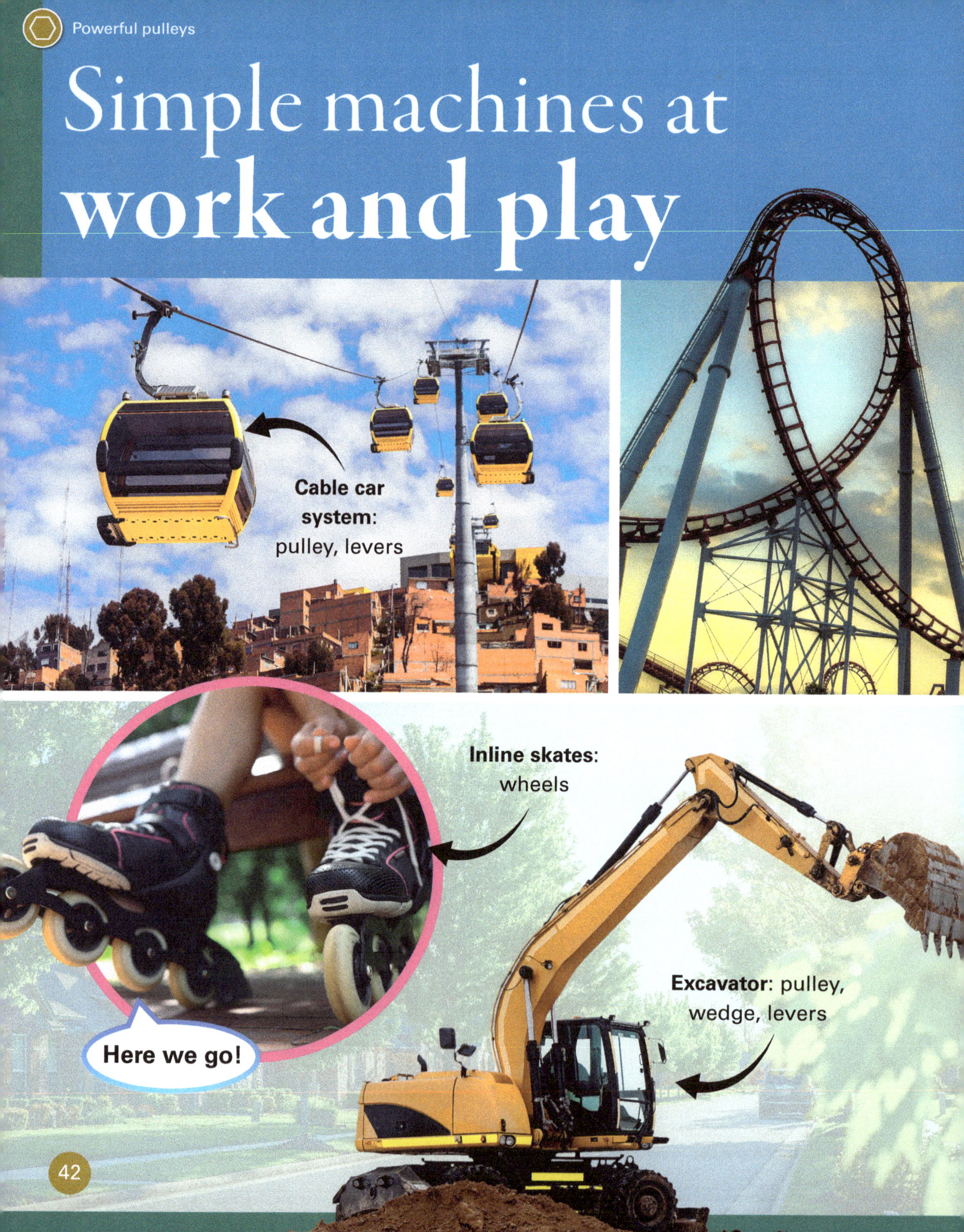

You've seen simple machines at work and play in all parts of our daily lives. Some help us do ***work*** by themselves. Other times one or more simple machines together create a compound machine. Keep your eyes open for simple machines everywhere!

Rube Goldberg was an inventor, engineer, author, and cartoonist born in 1883. He is best known for the machines he drew, complicated contraptions of chain reactions that accomplished a simple task in a ridiculous way. For example, he created a self-operating napkin contraption with 13 steps. That contraption included not only a live parrot and a rocket but also many simple machines!

Rube Goldberg machines

You will need:

The materials you use are up to you, though you definitely need a balled-up piece of paper and a container or trash can! Gather items that will make good parts for this contraption. You might also collect more items as you go.

Safety first!
Make sure an adult helps you with this activity.

Ideas for materials:

- Hot glue or other adhesive
- Cardboard (such as from cereal boxes)
- Scissors
- Marbles
- Paper towel tubes
- String
- Jumbo paper clips
- Rubber bands
- Other possible materials include dominoes, craft sticks, toy cars, marble run, straws, and a funnel.

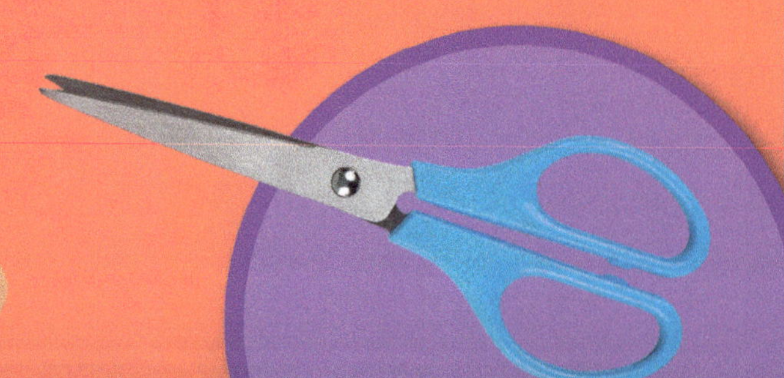

Give it a try

1. Research Rube Goldberg contraptions to learn more about these machines, see them in action, and get ideas.
2. Your task is to throw away the wad of paper. Design and draw a compound machine that accomplishes that task in several chain reaction steps. Try to include each simple machine in your design: inclined plane, lever, wedge, wheel and axle, screw, and pulley.
3. Gather the materials you want to use, and ask for adult help if needed.
4. Build! Assemble your contraption.
5. Test! After your contraption is complete, it's time to test it. Will it get the paper into the trash can?
6. Take notes when your compound machine is in action. What worked? What did not? Did your contraption complete the task?

Now that you have learned about simple machines, it's time to design and build your own compound machine. Your contraption will throw away a wad of paper using as many steps as possible and your knowledge of simple machines (you can skip the live birds and rockets!).

7. If your contraption did not complete the task, consider ways to adjust it. Perhaps you need to change the direction or strength of a force. Maybe you need to speed up or slow down an object in motion. Apply your knowledge of simple machines to make these adjustments.
8. Retest!
9. Share your contraption with family or friends. You could even have them join you in building the next contraption or adding steps to the one you built!

Try this next!

Rube Goldberg machines accomplish all kinds of ridiculous tasks. Redesign your machine to turn the page of a book, ring a bell, water a plant, or do any other task you dream up. Or you could add elements to your existing machine. Can you design a machine with 10 steps? 20?

Which of the simple machines were easiest to work with? Why do you think that is? Were there pairs of simple machines that worked best together in your contraption? What would you do differently in the design of your next contraption?

Index

Glossary

aqueduct (AAH kwuh dukt)—a large, raised channel used to move water

counterweight (KOWN tur wayt)—an equivalent force or weight

diameter (die AM uh tur)—the distance across a circle

effort (EHF uhrt)—the force applied to an object

force (FORS)—a push or pull that changes the speed or direction of an object, maintains the movement of an object, or keeps the object in place

friction (FRIK shuhn)—a force that resists movement between two objects touching each other

fulcrum (FUL krum)—the point on which a lever rests

gravity (GRAH vuh tee)—a force that pulls objects down to Earth's surface

input force (IN put FORS)—the force applied to a simple machine to move an object or load

load (LODE)—the object(s) that must be moved

mechanical advantage (meh CAN ih kul ad VAN tuj)—the advantage gained when simple machines are used in order to do work with less effort

output force (OWT put FORS)—the force applied to an object or load by a simple machine

work (WERK)—the amount of force (F) applied to an object to move it a distance (D); work (W) = F × D

www.ingramcontent.com/pod-product-compliance
Lightning Source LLC
LaVergne TN
LVHW060632110826
845147LV00014B/900

* 9 7 8 0 7 1 6 6 7 1 5 9 6 *